AF228843

Seahorses are small.

5

Seahorses live
in the sea!

7

Seahorses have tails.

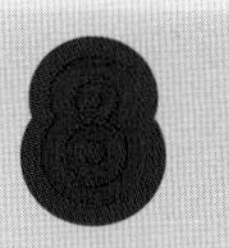

This seahorse is yellow.

11

This seahorse is red.

13

This is a baby
seahorse.

15

Seahorses swim.

Seahorses eat.

19

Seahorses have families.

21

Seahorses are fish!

23

Published in 2026 by The Rosen Publishing Group, Inc.
2544 Clinton Street, Buffalo, NY 14224

First Edition

Editor: Theresa Emminizer
Book Design: Jeffrey Taylor

Photo Credits: CoverJasper Hammink/iStock.com; p. 3 Aashu_b/Shutterstock.com; p. 5 JLImages/Alamy Stock Photo; p. 7 nickeverett1981/Shutterstock.com; p. 9 Amarphotocreater/Shutterstock.com; p. 11 Basicdog/Shutterstock.com; p. 13 Vojce/Shutterstock.com; p. 15 Rebecca Schreiner/Shutterstock.com; p. 17 GOLFX/Shutterstock.com; p. 19 Mike Workman/Shutterstock.com; p. 21 creativemarc/Shutterstock.com; p. 23 mc_pongsatorn/Shutterstock.com.

Cataloging-in-Publication Data

Names: Emminizer, Theresa.
Title: Seahorses / Theresa Emminizer.
Description: Buffalo, New York : PowerKids Press, 2026. | Series: Superstars of the sea Identifiers: ISBN 9781499451528 (pbk.) | ISBN 9781499451535 (library bound) | ISBN 9781499451542 (ebook)
Subjects: LCSH: Sea horses--Juvenile literature.
Classification: LCC QL638.S9 E46 2026 | DDC 597′.6798--dc23

Manufactured in the United States of America

Some of the images in this book illustrate individuals who are models. The depictions do not imply actual situations or events.

CPSIA Compliance Information: Batch #CSPK26. For further information contact Rosen Publishing at 1-800-237-9932.

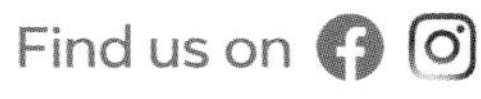

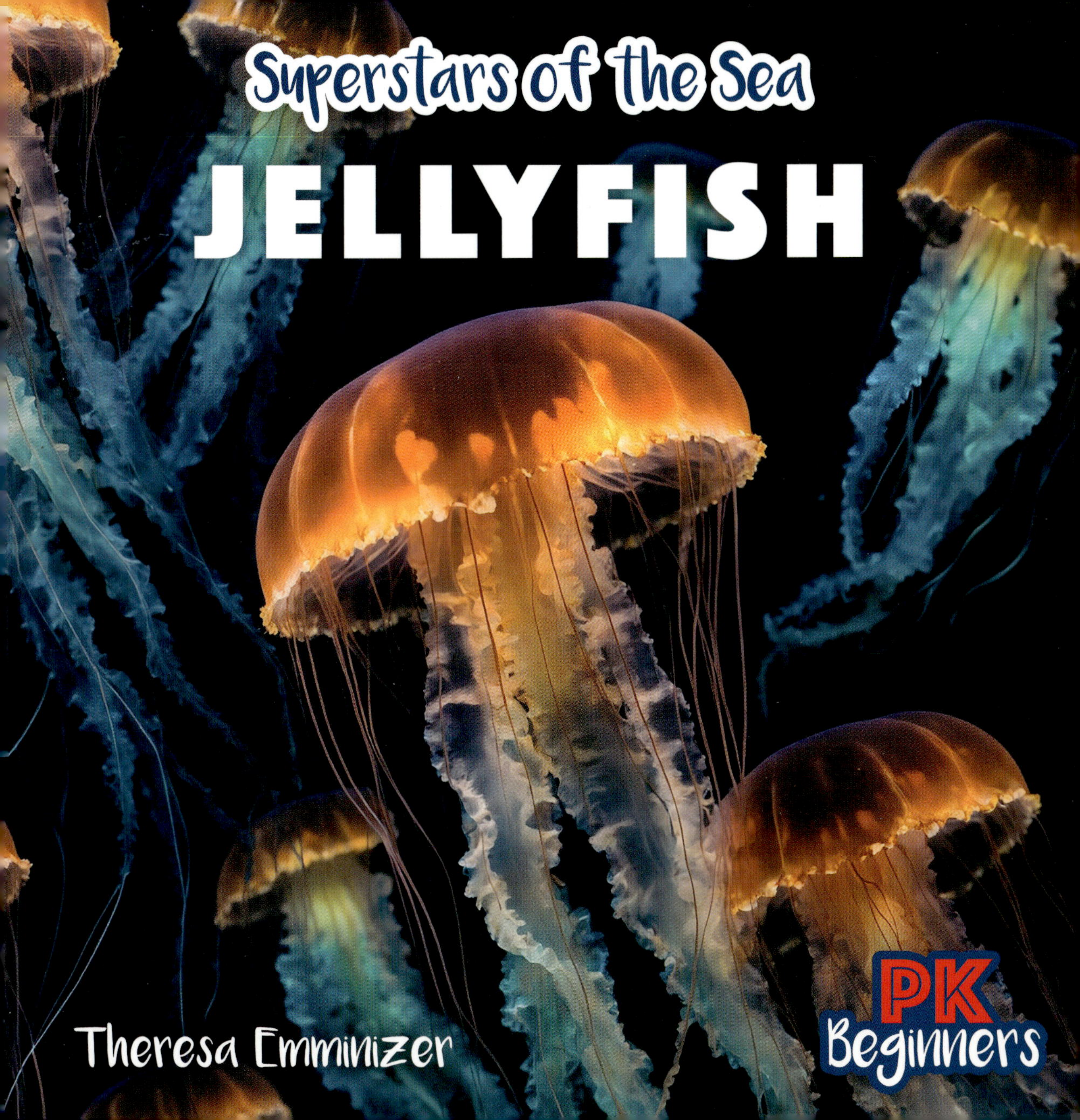

Superstars of the Sea
JELLYFISH
Theresa Emminizer
PK Beginners

Superstars of the Sea

JELLYFISH

Theresa Emminizer

PowerKiDS
press

PK
Beginners

Meet the jellyfish!